SATELLITE FEVER

by **MIKE PAINTER**

illustrated by **Mic Rolph**

*Making***sense***of***science**
Children's Books

series editor **Fran Balkwill**

Portland Press

Making **sense** *of* **science**

Children's Books

Other titles in the series

LIGHT UP YOUR LIFE by **David Phillips** • **PLANET OCEAN** by **Brian Bett**

THE SPACE PLACE by **Helen Sharman**

We wish to thank the following people who helped in the production of this book:
Editorial Advisor SUSAN DICKINSON
For Portland Press SOPHIE CAYGILL and ADAM MARSHALL
MATRA MARCONI SPACE
For Surrey Satellite Technology Ltd ED MILTON

First published in 1997 by Portland Press Ltd
59 Portland Place, London W1N 3AJ, UK

© text Mike Painter 1997
© illustrations Mic Rolph 1997

The author asserts the moral right to be identified as the author of the work

ISBN 1 85578 091 7   ISSN 1355 8560

Typeset by Portland Press Ltd
Originated and printed by Cambridge University Press, Cambridge, UK

Speeding around the Earth, high in space,
there are hundreds of man-made objects.

You are safer, healthier and wiser because of
them, and so is your planet.

This is a book about these very special
extra-terrestrial objects
SATELLITES.

There is one natural satellite that you can
see easily – the Moon.
It travels in space, around the Earth, following
a path we call its **orbit**.

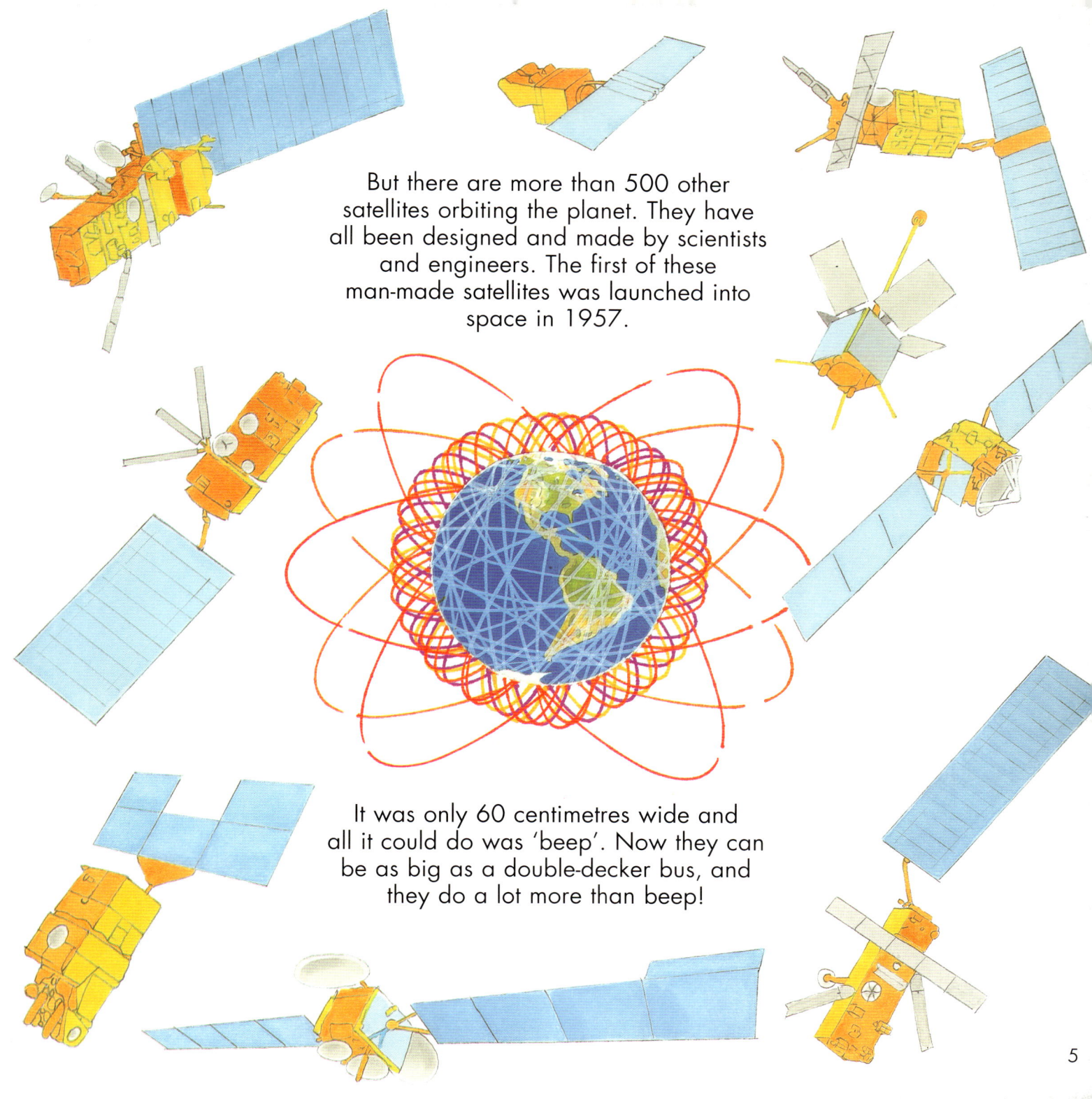

But there are more than 500 other satellites orbiting the planet. They have all been designed and made by scientists and engineers. The first of these man-made satellites was launched into space in 1957.

It was only 60 centimetres wide and all it could do was 'beep'. Now they can be as big as a double-decker bus, and they do a lot more than beep!

It is not very easy to get a satellite into space. A powerful rocket is needed to lift it against the invisible pull of Earth's **gravity**. Gravity is a force which holds everything down, people, cars, houses, skyscrapers, forests, even the great oceans and the air we breathe.

Because of gravity you can ski down a mountain, or fall off your bicycle. Without it, you and everything else would fly off into space!

The pull of gravity is strongest at sea-level. The further you go from the ground, the smaller the pull of gravity becomes.

The air you breathe is part of Earth's **atmosphere**, a mixture of gases and dust that protects the planet. On the ground, the atmosphere presses down on the Earth making **atmospheric pressure**, which you can measure with a barometer.

The further from the ground, the smaller the pressure, until eventually (at a height of over 250 kilometres) the barometer measures zero. This is a **vacuum**. There is no pressure here.

This is space.

Because there is no protective atmosphere in space, everything facing the Sun gets very hot – about 150 °C. Away from the Sun, in the deep blackness of space, the temperature can drop to −273 °C, the coldest anywhere in the Universe.

They must weigh as little as possible, otherwise a rocket would not be able to lift them.

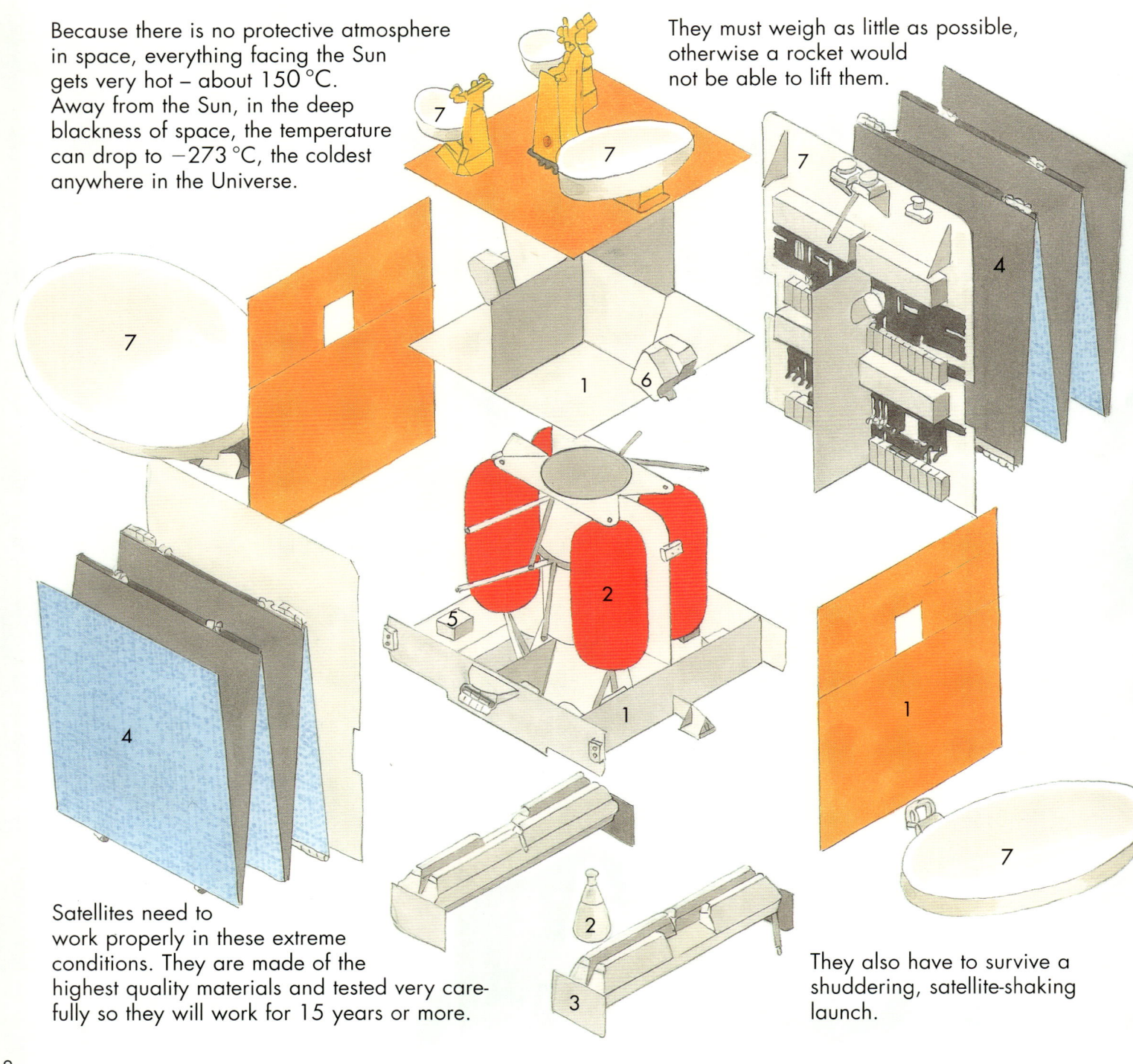

Satellites need to work properly in these extreme conditions. They are made of the highest quality materials and tested very carefully so they will work for 15 years or more.

They also have to survive a shuddering, satellite-shaking launch.

How to make a satellite

1 The first thing you need is a strong, but light, structure on which to bolt all the equipment.

2 Four tanks of chemical fuel, for the **propulsion system**, are attached. Pipes carry this fuel to one large and twelve small **thruster rockets** that can move the satellite about once it is in space.

3 Next come two batteries made of nickel plates in cylinders full of hydrogen.

4 These are charged by two large **solar panels** which turn sunlight into electricity.

5 A computer is connected to a receiver, to pick up radio commands from Earth. It is also connected to a transmitter that sends information back to Earth to confirm that everything is working correctly.

6 Two types of **sensor** are added to tell the computer the position of the Sun and Earth. This information is used to work out where the satellite is pointing and send signals to the thruster rockets to correct it.

7 Then the all-important **payload** is assembled. Each satellite is designed to do a particular job, called its **mission**. The payload, in this case electronic equipment and aerial dishes, allows the satellite to carry out its mission.

Finally, the satellite must be protected against the intense heat and cold of space. Surfaces that need to be kept cold are covered with mirrors to reflect away the Sun's rays. The rest of the satellite is wrapped in blankets to keep it warm. These blankets are made of layers of metal foil (like kitchen foil) separated by thin plastic.

Now the satellite is complete, but it must be tested fully to see if it will work in space.

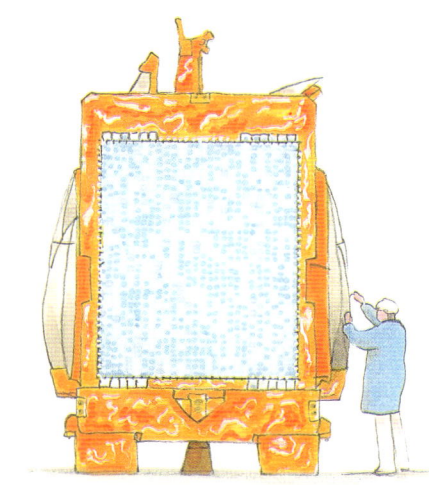

It is given a good shaking, to check that it will not disintegrate during the rocket launch. Then it is placed into a large vacuum chamber. It is made very hot, and then very cold, as would happen in space.

An extremely powerful rocket is needed to send a two tonne satellite speeding through the atmosphere and into space.

Most rockets have three stages, one mounted on top of the other. With the **nose-cone** added, this assembly will be as high as a 25 storey building.

The first level is really just a big fuel tank with the main engines at its base. It has **booster rockets** strapped to its side. The second stage is similar to the first, but shorter.

These two stages use about 260 tonnes of very dangerous chemical fuels, which ignite when they are mixed together at lift off. The fuel will burn away completely in five flame-blasting minutes.

The third stage carries eleven tonnes of hydrogen and oxygen as fuel. These are stored as liquid, and have to be kept at a very low temperature, otherwise they will boil away before take off.

At the top of the third stage is the 'brain' of the rocket. Its computers and guidance equipment make sure the rocket follows the right course into space and reaches the correct orbit.

Finally, right at the top of the rocket is our satellite, folded up to fit in a streamlined nose-cone.

The rocket weighs about 420 tonnes. Over half of this is fuel.

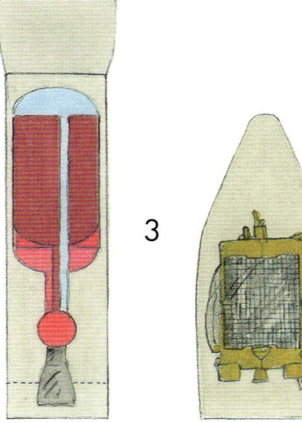

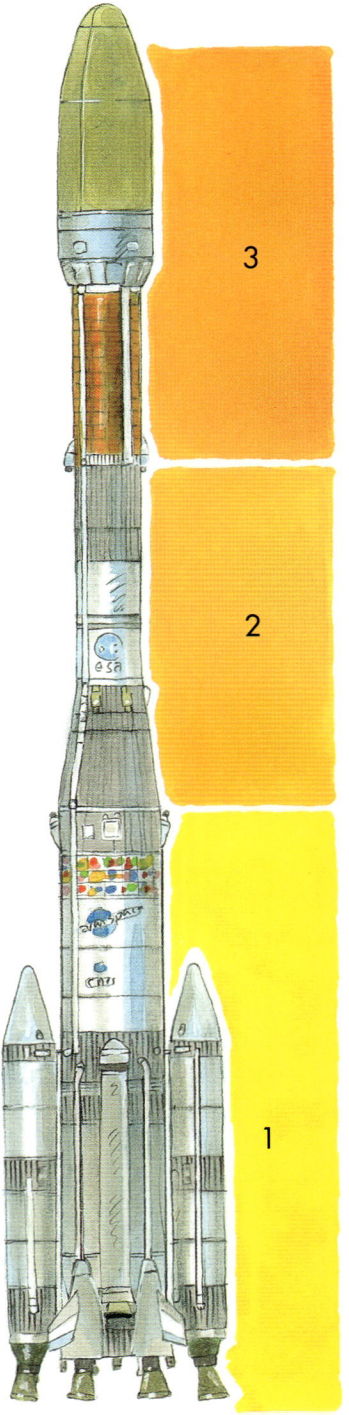

The satellite is now ready to be taken to the launch site, which is usually a journey of thousands of kilometres. The precious cargo, with its solar panels and aerial dishes already folded, is loaded into a large container and transported there by aeroplane or by train.

When it arrives, the satellite is tested all over again. This is the very last chance to make sure everything is in working order.

The satellite is loaded gently on to the top of the rocket and the protective nose-cone is attached. This streamlines the rocket and protects the satellite from the battering it gets as the rocket forces its way through the atmosphere.

When everything is ready, and weather conditions are OK, the countdown starts...

10 9 8 7 6 5 4 3 2 1

IGNITION!

As it slowly lifts, the engines push against gravity with a force of 560 tonnes. Even so, the huge rocket moves quite slowly. In the first few seconds it moves less than two metres. But the rocket keeps fighting against gravity.

LIFT OFF!

The rocket starts to rise, through huge clouds of dust, smoke and flame.

After one minute, the rocket has reached a speed of 950 kilometres per hour (kph), the top speed of a jumbo-jet. The on-board computer sends a command signal to release the booster rockets. Three and a half minutes later, the rocket has reached a height of 65 kilometres and is travelling at over 8000 kph.

The first stage quickly uses up all its fuel and is commanded by the computer to fall away. The rocket races on under the power of the second stage and, at a height of 110 kilometres, leaves the Earth's atmosphere. The buffeting stops at last. There is no need for the protective nose-cone now. A signal is sent by the rocket's computer, and the nose-cone is split in half from top to bottom and falls away.

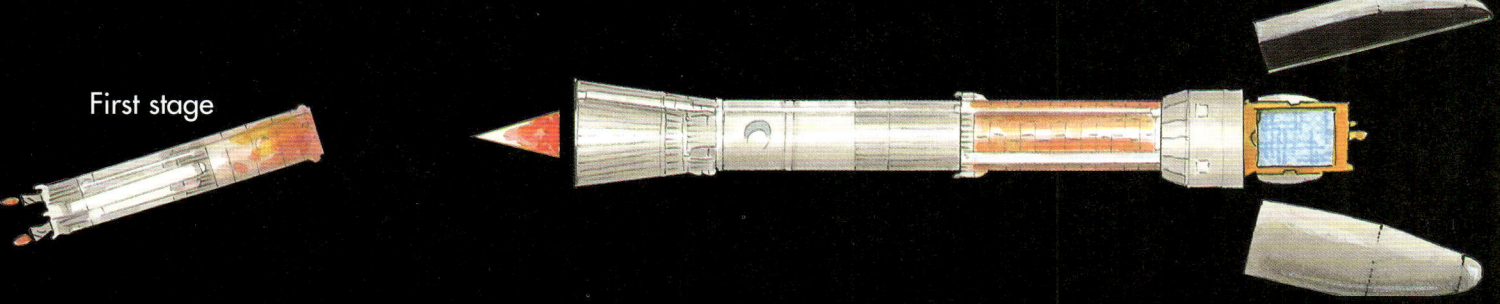

First stage

Five and a half minutes after lift off, the second stage is exhausted and it too is instructed to separate from the rocket.

CLUNK !

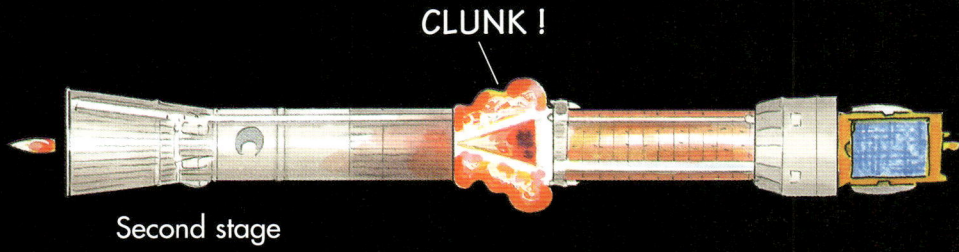

Second stage

The third stage now takes over, carrying the satellite into its orbit. Twenty minutes after lift off, and at a height of 300 kilometres, the satellite is injected into orbit at a speed of about 35,000 kph – faster than a bullet from a gun!

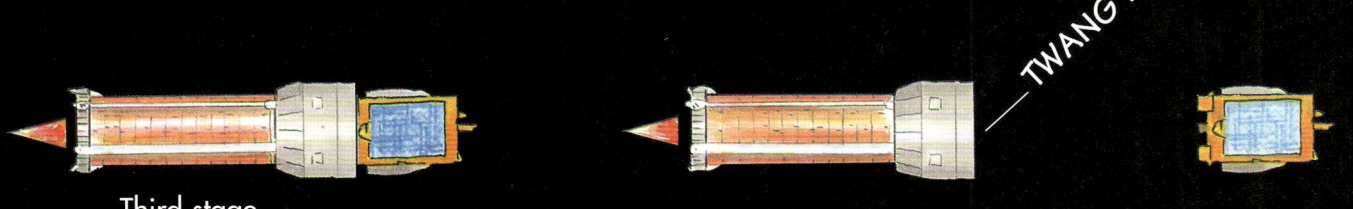

TWANG !!

Third stage

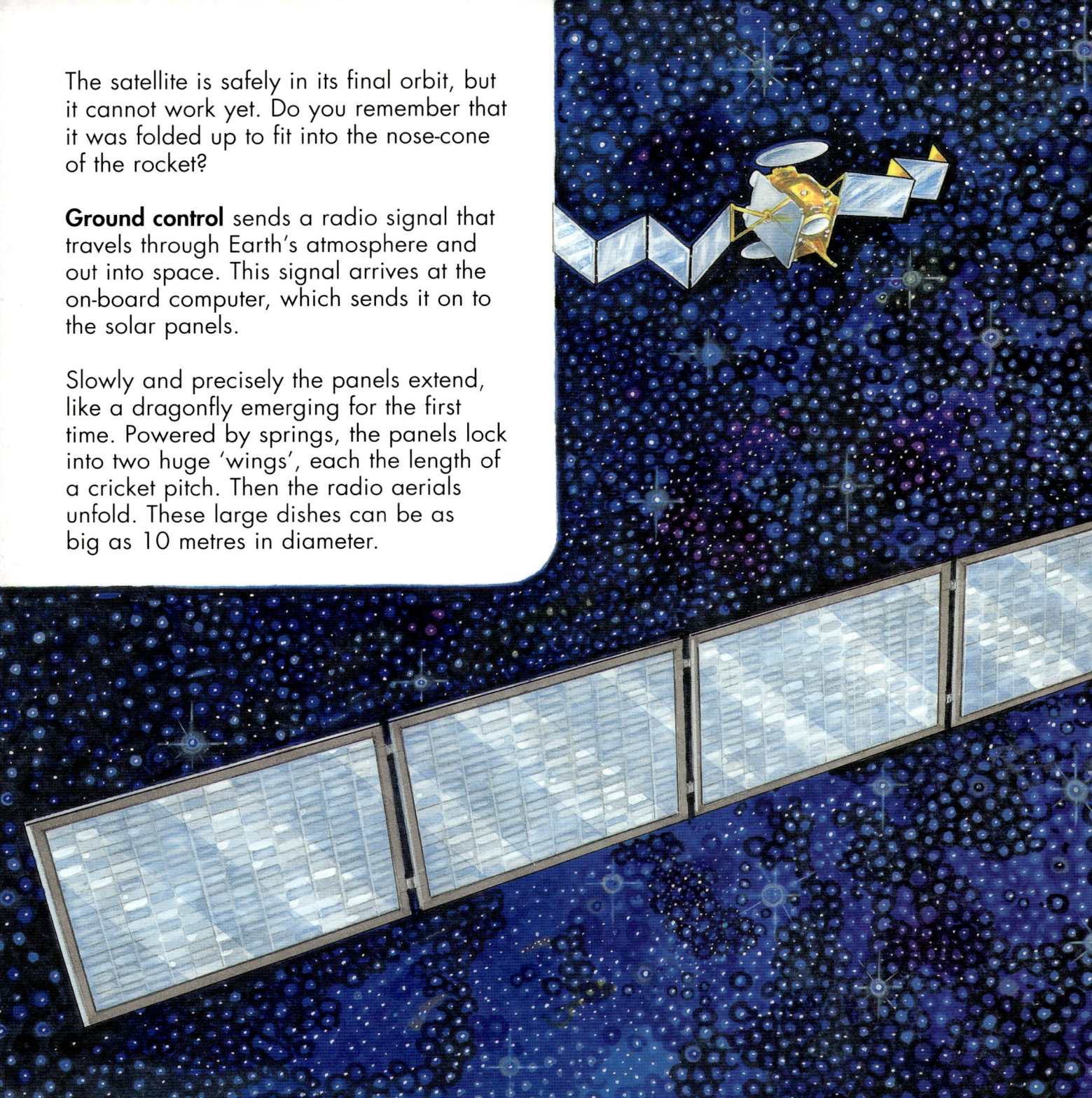

The satellite is safely in its final orbit, but it cannot work yet. Do you remember that it was folded up to fit into the nose-cone of the rocket?

Ground control sends a radio signal that travels through Earth's atmosphere and out into space. This signal arrives at the on-board computer, which sends it on to the solar panels.

Slowly and precisely the panels extend, like a dragonfly emerging for the first time. Powered by springs, the panels lock into two huge 'wings', each the length of a cricket pitch. Then the radio aerials unfold. These large dishes can be as big as 10 metres in diameter.

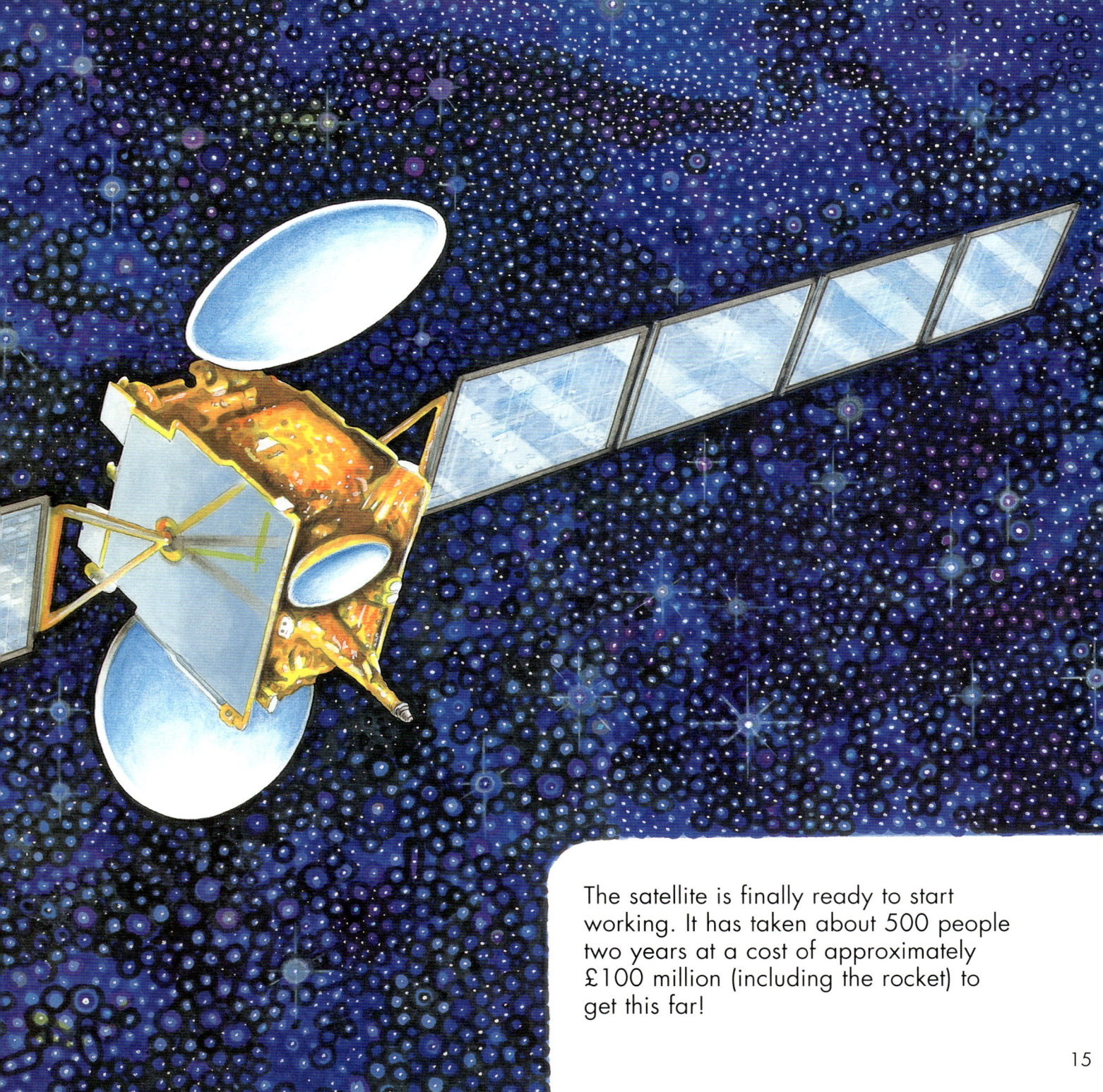

The satellite is finally ready to start working. It has taken about 500 people two years at a cost of approximately £100 million (including the rocket) to get this far!

What are satellites doing as they orbit around our planet? Remember, each satellite is designed to do a particular job, called its **mission**.

The satellite which has just been launched is a **communications satellite**, and its mission is to broadcast television programmes from its position high above Earth.

Television signals are broadcast as **microwaves**. These are invisible waves that behave like light waves.

Microwaves travel at the same incredible speed as visible light (300 million metres per second). But, unlike visible light, they can pass through buildings and other obstacles on their way to your television.

Microwaves travel in straight lines, so people who live over the horizon or in a deep valley cannot 'see' a transmitter and are therefore unable to receive a signal. That is why television transmitter aerials are placed on top of masts or towers built on high ground. Then they are able to 'see' further and transmit television over an area within a radius of about 80 kilometres.

If a television transmitter and an aerial
are fitted to a satellite 36,000 kilometres
above Earth, the satellite is so high that it
can 'see' one-third of Earth's surface. To
receive the programmes, your television
set must have a dish aerial pointed at the
satellite, and the satellite must always
remain in line with that dish.

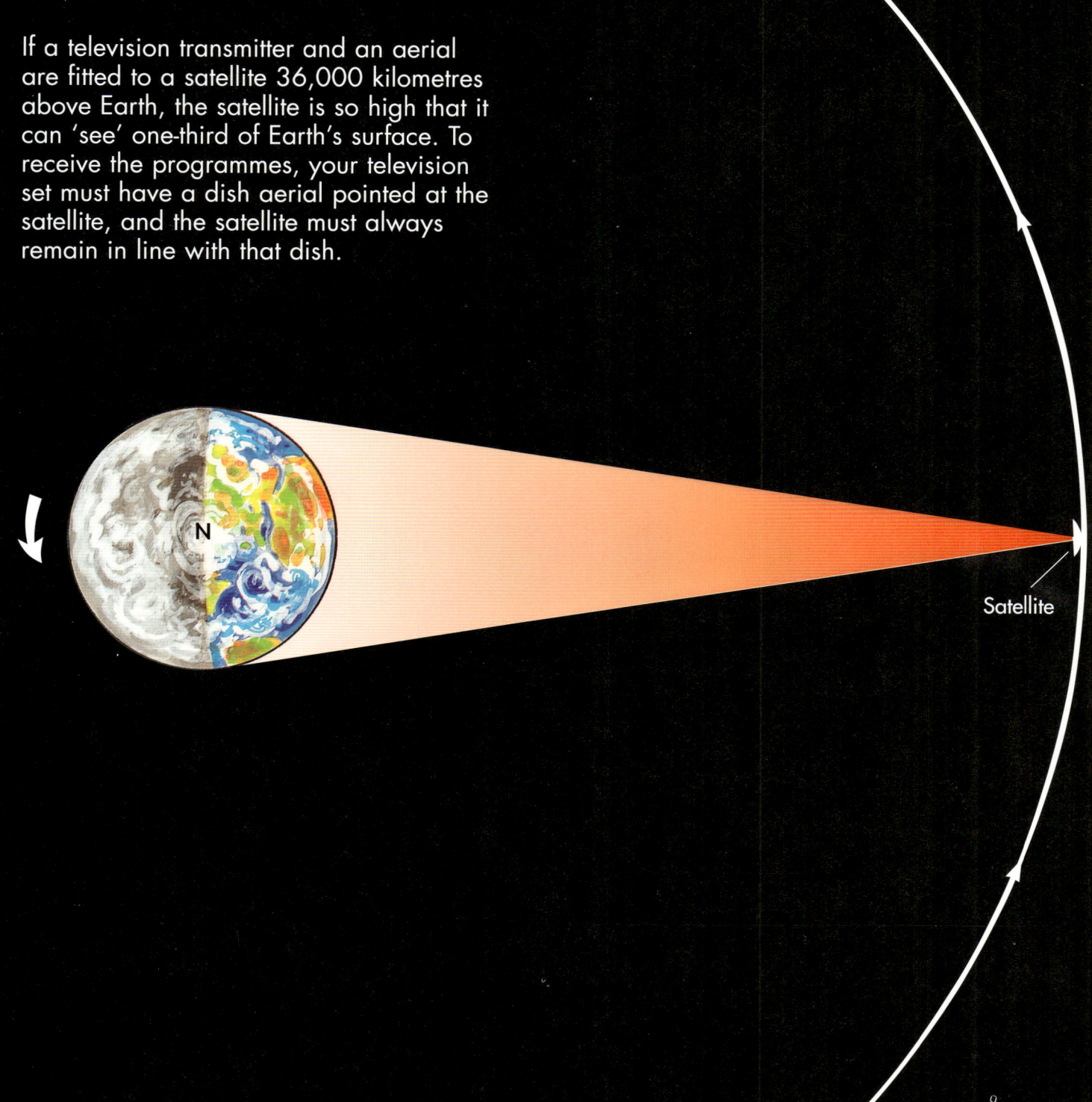

Satellite

To keep the satellite at the same place in the sky, a special orbit is used. A satellite travelling 36,000 kilometres above the equator takes exactly 24 hours, or one day, to orbit the Earth. This is the same time it takes for the Earth to make one complete rotation.

Because communications satellites appear to be stationary in the sky, the ground stations and your home receiver dish do not have to track a moving satellite to catch its signals.

If the satellite is moving in the same direction as the Earth is turning, it will appear to 'hover' over the same place for the whole of its life in space.

This is called the **geostationary orbit**.

The dishes are fixed in one direction and 'see' their satellite all the time, just as an ordinary radio or television aerial can 'see' the transmitter tower. Next time you walk down your street, look which way the dishes are pointing.

One communications satellite can 'see' one-third of the Earth's surface, so it can send microwave signals to any receiver in that whole area. Three satellites spaced equally around the globe are able to cover the whole world.

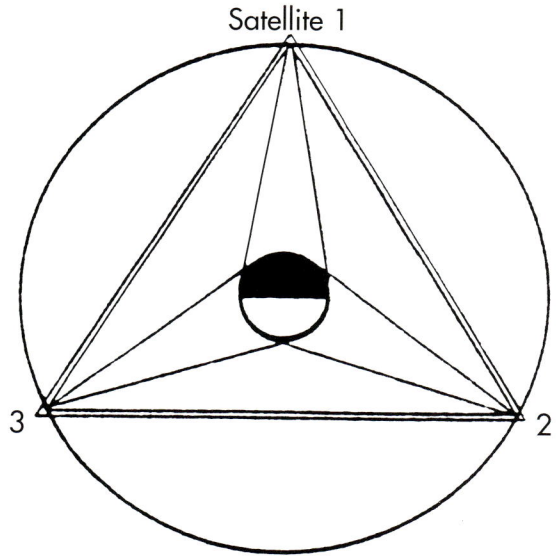

Diagram from Arthur C. Clarke's original 1945 memo to the Council of the British Interplanetary Society explaining the basic principles of geostationary communications satellites

This means that satellites can carry live television signals from one side of the Earth to the other. Just one satellite will soon be able to broadcast 100 different television channels into your home. Other satellites carry telephone and fax messages, and connect computers together.

Before very long you will be able to telephone home via a satellite from the remotest parts of the world, using a pocket-sized mobile phone.

Satellites can even send telephone messages to ships, yachts, aircraft and trucks. This is a difficult job because the terminals they transmit to are small, moving and hard to 'see', especially during a violent storm at sea.

n the same orbit as our communications satellite, there are satellites which carry a camera to take pictures of the Earth, especially its clouds. These pictures are essential to weather forecasters. Satellites track hurricanes, tornadoes, typhoons, erupting volcanoes and oil slicks, giving vital warnings of approaching disasters.

Weather forecaster checks satellite pictures
Storm brewing

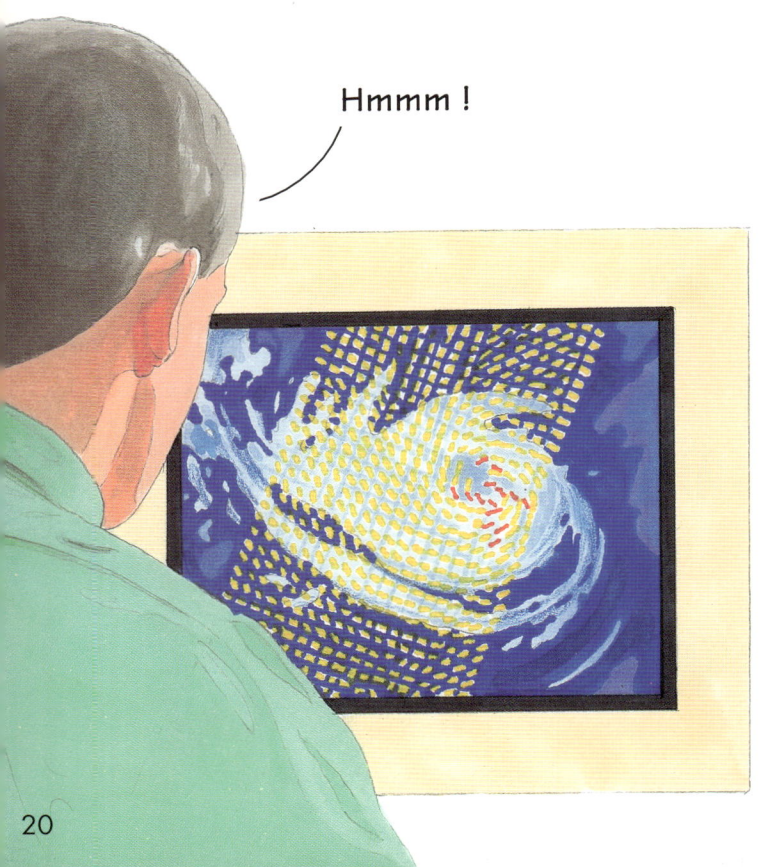

Hmmm !

24 hours later

Storm developing into hurricane
A forecast is made...

Hurricane will hit coast in about 12 hours time and that will coincide with high tide

EMERGENCY WARNING

**Alert emergency services...
...Evacuate schools and hospitals
Board up windows...Warn other coastal areas...Set up satellite communications base...**

Hurricane STRIKES!

Satellite communications call in more emergency help... World wants to know what is happening...Pictures sent by satellite show devastation...

Thanks to satellites and weather forecasts, injuries have been kept to a minimum.

Earth Observation satellites carry a camera along an orbit that passes over the North Pole, down over the South Pole, and back north again. The camera continuously takes pictures of Earth, which are sent back to receiving stations on the ground.

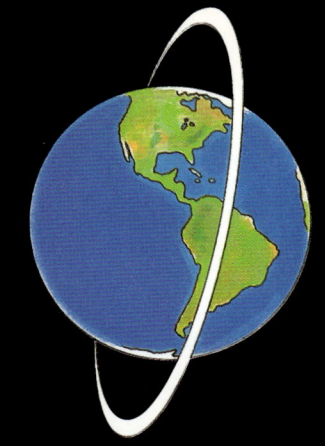

So that the camera can take really good pictures, these satellites are only 800 kilometres high. From this height, the camera 'sees' one strip of the Earth every 96 minutes, as it follows its orbit. The planet turns beneath the satellite. With every new orbit the camera 'sees' a new strip, which runs alongside the strip it has just photographed. In this way, the camera takes pictures of the whole planet, and the satellite transmits them back to Earth.

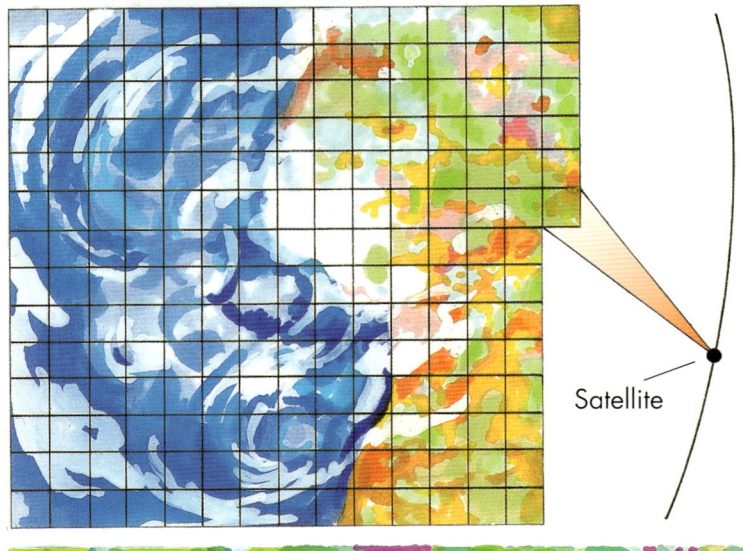

Satellite

In the same orbit there are satellites that carry a **radar** payload instead of a camera. Radar pictures can be taken through clouds and at night. Other instruments on the satellite can measure gases in the atmosphere and the temperature of seas and oceans.

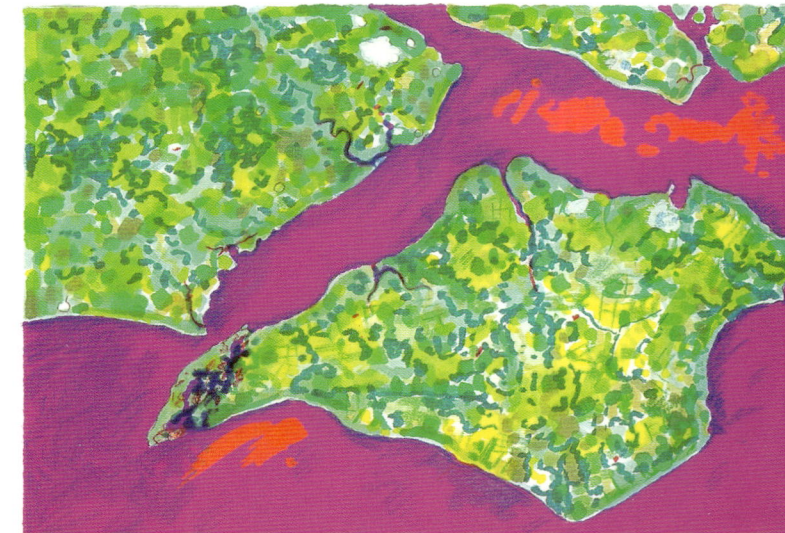

These are the satellites that take care of you and your planet. The pictures they take help with weather forecasts and making maps. They are used to watch crops, search for oil and precious minerals, spot dangerous icebergs and monitor floods. Even small movements of the Earth's surface can be measured, which may help to predict earthquakes.

Most important of all, the satellites help track down pollution. Some scientists think that our world is getting warmer because of pollution and because we are losing the layer of ozone that protects us from the Sun's radiation. Earth Observation satellites are vital to research into global warming.

A satellite picture of the hole in the ozone layer above the South Pole

If you look carefully at the night sky, you might see one of these satellites. Like tiny shooting stars, they move across the horizon, visible because their shiny surfaces reflect the Sun's rays.

For thousands of years explorers and sailors have relied upon the stars to find their way. Now there is a constellation of 24 satellites arranged evenly around the Earth, about 19,000 kilometres up. They act as **navigation beacons**, continuously sending out radio signals that tell their position in space. Anyone with a special receiver can pick up the signals.

The receivers are small enough for walkers and explorers to carry. They can even tell mountaineers how high they have climbed. Ship and aircraft navigators depend on these satellite beacons.

You cannot get lost if you are in touch with navigation satellites.

NORTH POLE
Fish Restaurant

But he wants a BURGER !

I must be lost !!

Armed forces depend on satellites to fight a modern war and to keep the peace. In 1989, over 100 satellites were used in the Gulf War. In fact, this is sometimes called the first 'Space War'.

Secure communications can be set up, anywhere in the world, within an hour. Navigating by satellite means everyone knows exactly where they are and which way to go. Accurate weather forecasts help decide the best time to attack. Spying on the enemy is easy when satellites are listening to their radio messages and taking photographs of targets or bomb damage.

Missiles can be detected by satellite and intercepted before reaching their target. Television satellites send pictures from war zones all over the world, and broadcast programmes from home for the soldiers.

Satellites also keep the peace. When a war is over, they restore communications quickly so that relief operations can begin. Spy satellites can help to make sure everyone keeps to the peace agreement. Countries with satellites have the power to watch closely those making preparations for war. This can help to make a peaceful planet.

Satellite terminals are coloured red

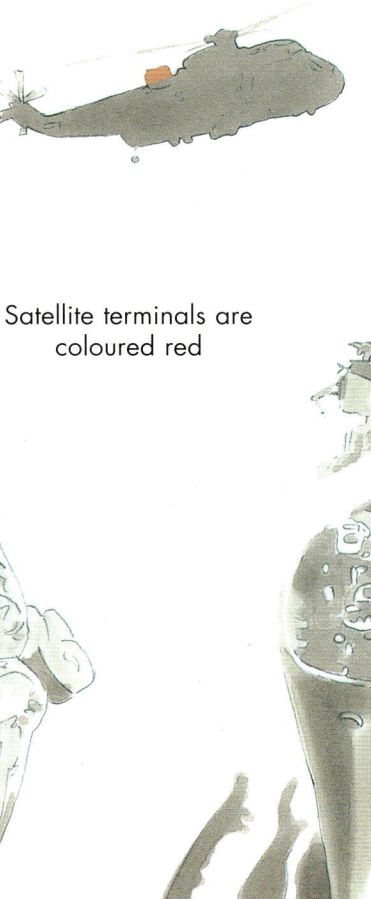

How do all these satellites stay speeding around Earth in a precise orbit for years? Why don't they crash into one another? Why don't they fly off into space?

They stay in place because gravity is pulling them towards Earth. One way to understand this is to think of a ball attached to a piece of string. Imagine the ball is a satellite and your hand is the Earth. If you swing the ball around fast enough, it 'orbits' around your hand. Then you can feel a force that tries to make the ball fly off into 'space'. This is called **centripetal force**. The string stops the ball from flying off. The string is holding the ball back just as the pull of gravity holds the satellite in its orbit. The faster the ball moves around your hand, the greater the centripetal force.

At least he didn't use ME !

When the pull of gravity downwards is equal to the centripetal force of the satellite speeding around the Earth, the satellite stays in its orbit. Satellites in low orbit have to travel very fast, at about 27,000 kph, because the pull of Earth's gravity is strong. In higher orbits, the pull of gravity is less, so satellites can travel more slowly. Sometimes the orbit height of a satellite has to change, for instance when it stops working.

An old communications satellite at 36,000 kilometres is sent to a higher 'graveyard' orbit to make room for a replacement. Its speed is increased for a while so that it can escape further from the pull of gravity. Earth Observation satellites at a height of 800 kilometres are slowed down so that they re-enter the atmosphere, where the **friction** of the air makes them burn away completely.

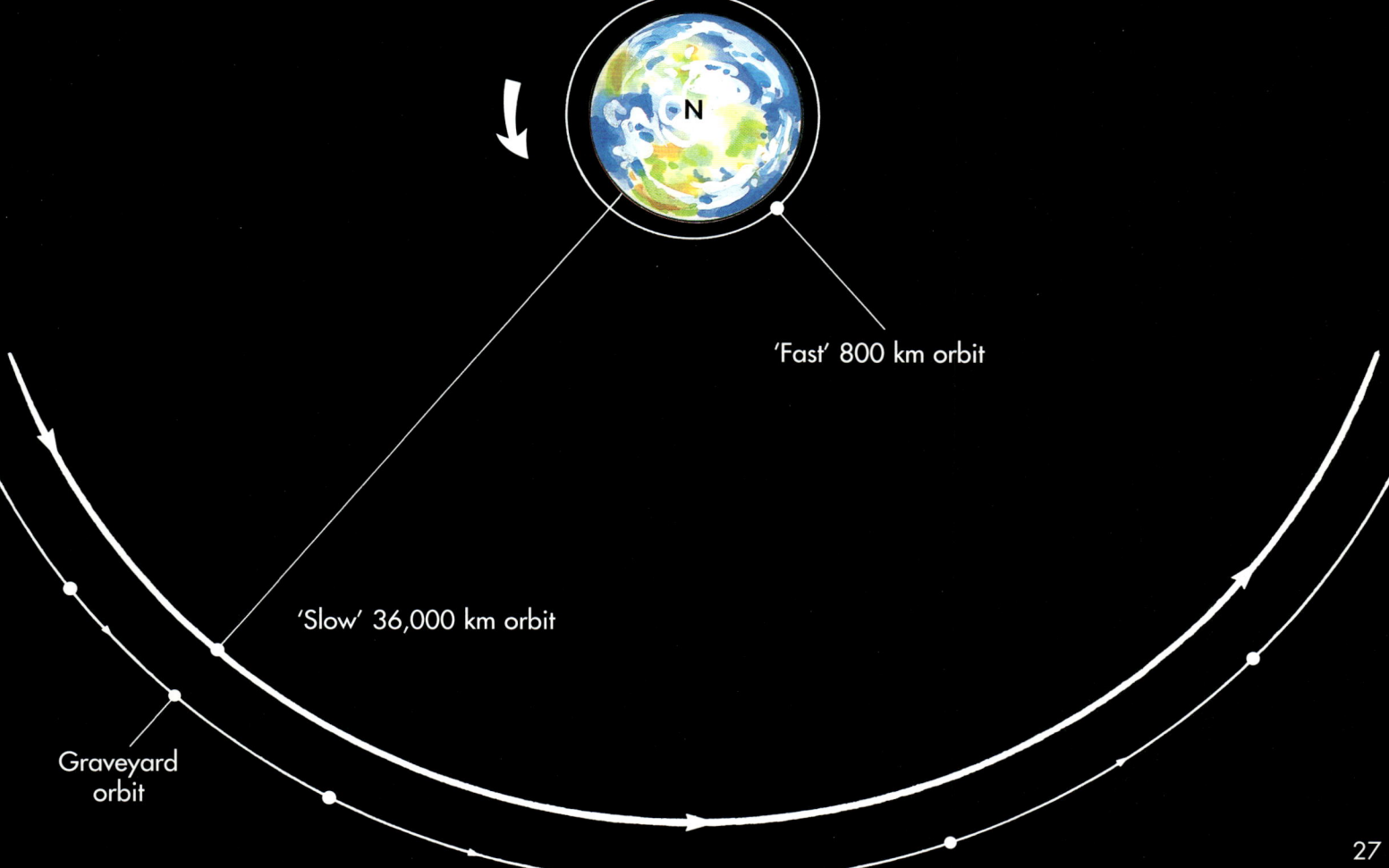

'Fast' 800 km orbit

'Slow' 36,000 km orbit

Graveyard orbit

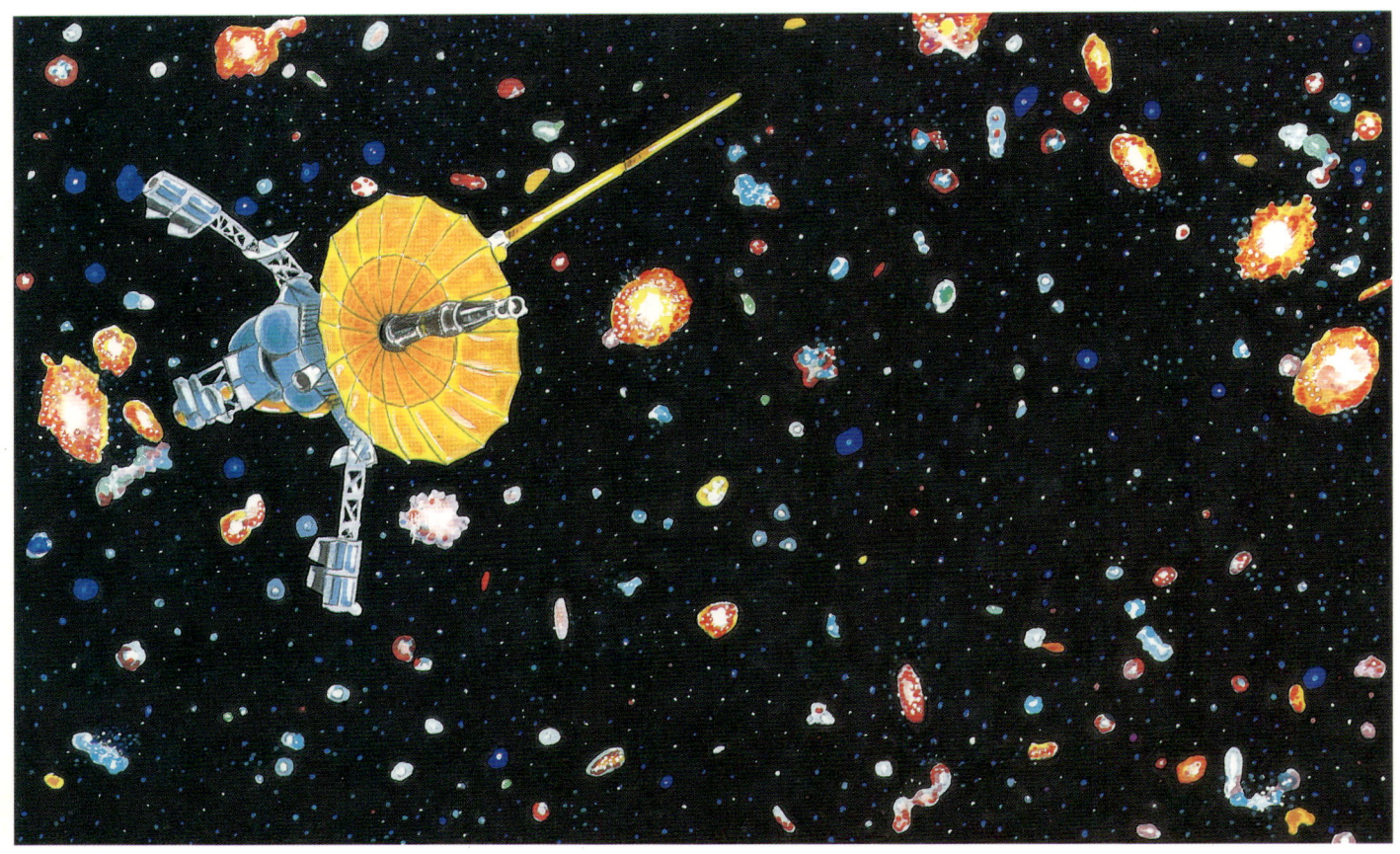

Our Sun is a star and Earth is one of its planets. Scientists think there may be as many as one thousand million stars in our galaxy alone, each surrounded by its planets. And there are about one hundred thousand million galaxies just like ours. It is almost impossible to understand anything so vast. There are so many questions to answer. Exactly how big is the Universe? How did it all begin? Is there life on other planets? Can the massive explosions of energy that can be seen through a telescope help us find ways to make cheaper electricity? We can begin to answer these questions, and many more, by using satellites.

Not all spacecraft become satellites in Earth orbit. They can be sent on missions to the Solar System and beyond. Controlled by microwave command signals from Earth, they can orbit other planets or investigate stars and comets.

A spacecraft has been launched to observe the Sun from a position one and a half million kilometres from Earth. Scientists hope to understand how the Sun controls our climate and the space between the planets of our solar system. Missions have already been sent to Mars and Jupiter, and in 2001 a spacecraft will approach Saturn and go into orbit around its rings. A **probe** will descend by parachute on to Titan, one of Saturn's moons, taking measurements on the way. Scientists think that Titan's atmosphere is similar to that of the early Earth.

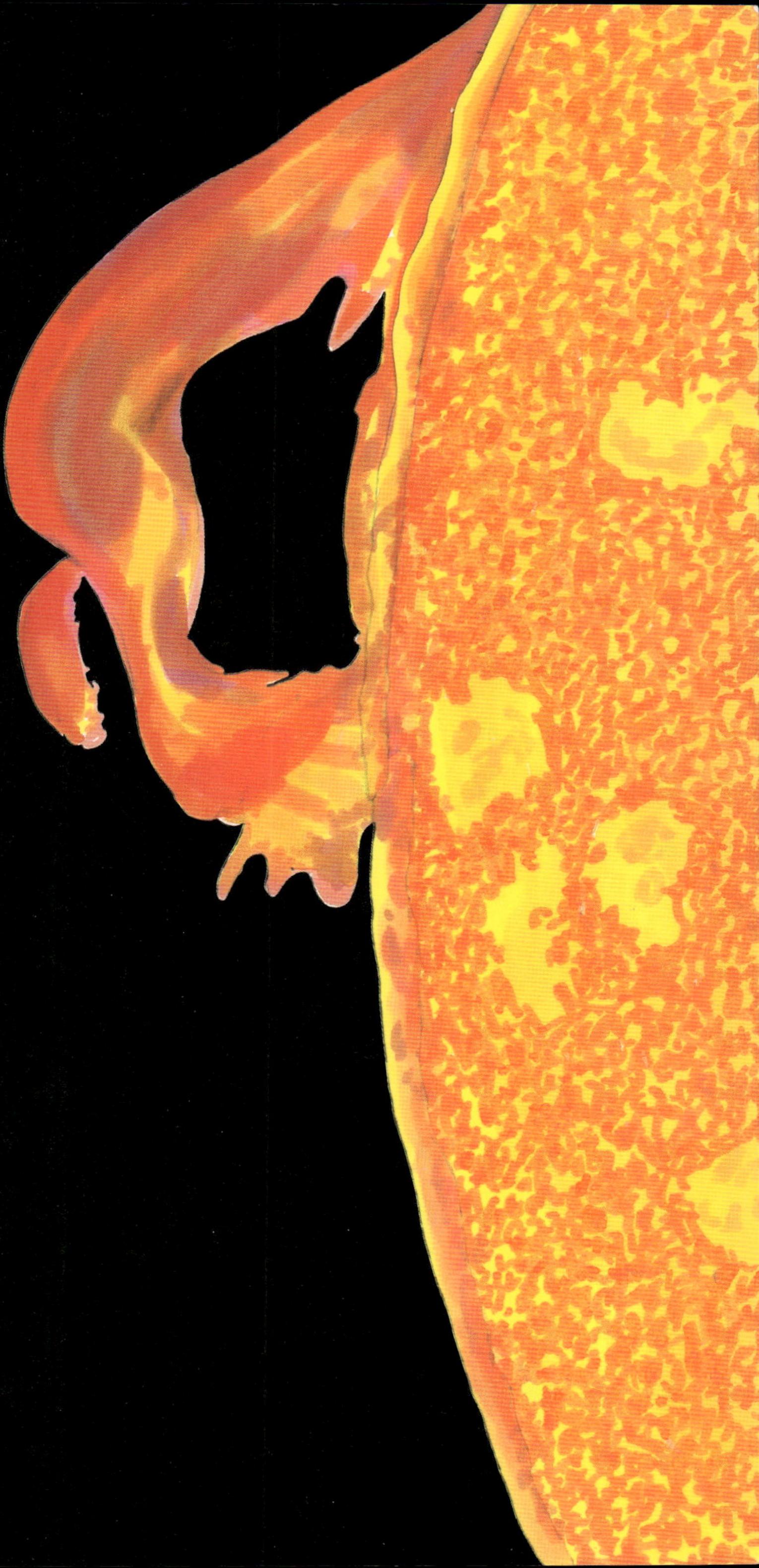

Another spacecraft will catch up with a comet called Wirtanen which is 860 million kilometres away, and is made of material similar to that of the early Solar System. A microwave signal travelling at 300 million metres per second will take 48 minutes to travel from Wirtanen to Earth!

Astronomers use telescopes to see beyond the Solar System to the farthest parts of the Universe. But Earth's atmosphere distorts the light from distant galaxies. Twinkling stars are pretty, but they cannot be seen clearly through a telescope on Earth. They appear to wobble!

An exciting way to solve this problem is to carry the telescope on a satellite and transmit the pictures back to Earth. Such a telescope can still be steered towards the stars by signals sent from a ground station.

The biggest and most powerful telescope in orbit is the **Hubble telescope**, 600 kilometres above Earth. It did not work properly when it was first launched by the Space Shuttle. The pictures were not as clear as expected and astronauts were sent to replace faulty equipment.

Now the telescope is sending back the most exciting pictures from the very edge of the Universe, helping us to understand how it all began. It is as if astronomers had been looking through dark glasses and now they have powerful binoculars.

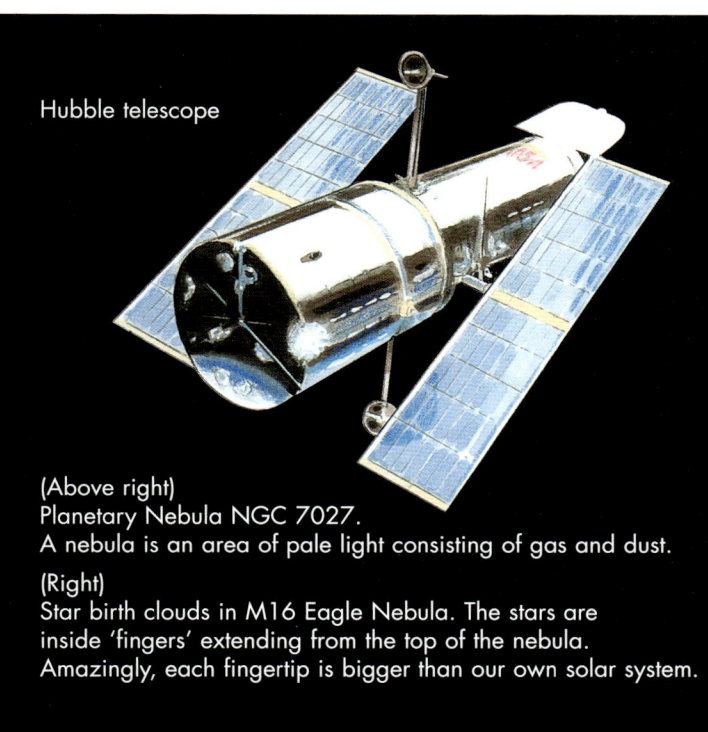

Hubble telescope

(Above right)
Planetary Nebula NGC 7027.
A nebula is an area of pale light consisting of gas and dust.

(Right)
Star birth clouds in M16 Eagle Nebula. The stars are inside 'fingers' extending from the top of the nebula. Amazingly, each fingertip is bigger than our own solar system.

Imagine that, instead of building a new rocket
for each satellite launch, we were able to use
a 'space plane'. It would take off like an
ordinary aircraft, but keep on climbing until
it left the atmosphere and went into orbit.
The cargo doors would open, the satellite
would be released and the space plane would
return to land at an airport and prepare for
the next launch.

The same plane could carry tourists into space
to stay at hotels built like a space station.
Imagine booking your holiday in space!
The launch would feel like a ride on a roller
coaster at a fairground, only longer. From
your hotel window you could watch
satellites being launched, at least one every
day, because by then there would truly be

Satellite Fever!